Introducing Logic

Introducing Logic

George N. Frempong

Contents

Preface

The aim of this book is to help students gain proficiency in correct logical reasoning.

The book has been developed from materials used in teaching mathematics at Accra High School, where taught for 30 years. I have tried to present the concept in a form that is easy to understand. The exercises at the end of each chapter are varied, providing basic skills and the understanding of the concepts. The clarity of the text makes the book excellent for self-study.

1 Statements and Connectives

There are several reasons for studying logic. Logic enables you to think clearly, communicate effectively and make more convincing arguments. In addition, logic enable us gain proficiency in correct mathematical reasoning.

Statements

In our daily interaction, we communicate ideas in writing or in speech by using sentences. Consider the following sentences.

1. Dr Kwame Nkrumah was the first president of Ghana.

2. The largest city in Togo is Accra.

3. Ama celebrated her fifth birthday yesterday.

These sentences express ideas that are either true or false. A sentence that is either true or false is called a statement. Statement 1 is true because Dr Kwame Nkrumah was the first president of Ghana. Statement 2 is false because Accra is a city in Ghana. We cannot determine whether Statement 3 is true or false, but definitely it should one or the other.

Simple Statements

A statement that conveys one idea is called a simple statement. Examples include "Kofi is a college student" and "River Volta is in Ghana".

Compound Statements

Consider the statement "Esi is a High School student and Esi likes reading". This statement is obtained by joining the two simple statements "Esi is a High School student" and "Esi likes reading" by the word "and".

Statements consisting of two or more simple statements are called compound statements. In addition, we consider a simple statement that has been negated to be a compound statement. The word not is generally used to negate a statement.

2   Statements and Connectives

Connectives

We use words such as

and,   or,   if … then …,   if and only if

to join two or more simple statements. In logic, we call these words connectives. Although not modifies a simple statement we consider the word not to be a connective.

In the study of logic, it is customary to represent each statement with a lower-case letter such as p, q and r, and each connective with a symbol. Table 1 shows the connectives and their symbols.

Table 1

Connective	Symbol
Not	$\sim$
Or	$\vee$
And	$\wedge$
If … then	$\rightarrow$
If and only if	$\leftrightarrow$

Each connective is a very important part of a compound statement because the truth value of a compound statement depends not only on the truth value of its components but also on the meaning we give to the connectives.

Types of Statements

Compound statements are given specific names depending on the connective used.

Negation

Sometimes it is necessary to change a statement to its opposite meaning. To do so, we use the negation of the statement. For example, the negation of the statement "Kofi is a college student" is "Kofi is not a college student". If p represents the simple statement "Kofi is a college student" then $\sim$ p represents the compound statement "Kofi is not a college student".

Note that the negation of "The water is hot" is not "The water is cold". If something is not hot, it is not necessarily cold. The correct negation is "The water is not hot".

By convention letters such as p, q and r are used to represent statements that are not negated. We represent negated statement with the negation symbol ~. For example, consider the statement "Afua is not at school." The word not indicates that it is a negation. To write this statement symbolically, we let p represent Afua is at school. Then ~p would represent the statement "Afua is not at school."

An important point to remember is that a negation symbol has the effect of negating only the statement that directly follows it. To negate a compound statement, we must use parentheses. When the symbol ~ is placed in front of a statement in parentheses, it negates the entire statement.

Conjunction

A conjunction is a compound statement formed by joining two or more statements with the word "and".

If p and q are two simple statements, then the conjunction p and q is denoted symbolically by p ∧ q. This is read p and q.

Examples

Write the following conjunction in symbolic form.

1. Ama is at home and she is watching television.

    Let p and q represent the simple statement

    p: Ama is at home.

    q: Ama is watching television.

In symbolic form the compound statement is p ∧ q.

2. Adjoa is in the kitchen but she is not cooking

    Let p and q represent the simple statements.

    p: Adjoa is in the kitchen.

    q: Adjoa is cooking.

    In symbolic form, the compound statement is p ∧ ~ q.

4 Statements and Connectives

Disjunction

A disjunction is a compound statement formed by joining two statements with the word "or".

If p and q are two simple statements, the disjunction p or q is denoted symbolically by p ∨ q. This is read p or q.

Examples

Write the following disjunction in symbolic form.

1. Ama is home or Ama is in the office.

   Let p and q represent the simple statements.

   p: Ama is home.

   q: Ama is in the office.

   In symbolic form, the compound statement is p ∨ q.

In the statement Ama is home or Ama is in the office we interpret the word or as meaning either Ama is home or she is in the office. The word or in this case is the exclusive or. When the exclusive or is used one or the other of the events can take place, but not both.

2. Kwame will study or he will listen to music.

   Let p and q represent the simple statements.

   p: Kwame will study.

   q: Kwame will listen to music.

In symbolical form the compound statement is p ∨ q.

In the statement Kwame will study or he will listen to music we interpret the word or as meaning Kwame will study or Kwame will listen to music or that Kwame will study and he will listen to music. The word or in this example is the inclusive or. When the inclusive or is used, one or the other, or both events can take place. In logic the word or in a statement will mean the inclusive or unless stated otherwise.

Conditional

A conditional statement is a statement that can be written in the form, if …then. If p and q are two statements, then the conditional statement is denoted by the symbol p → q, read if p then q. The statement immediately following if is called the hypothesis, and the statement following then is called the conclusion.

Examples

Write the following conditional in symbolic form.

1. If the sun is shining, then I will play outside.

    Let p and q represent the statements:

    p: The sun is shining.

    q: I will play outside.

    In the symbolic form, the statement is p → q.

2. If she smiles, then she is happy.

    Let p and q represent the statements:

    p: She smiles

    q: She is happy.

    In the symbolic form, the statement is p → q.

Biconditional

A biconditional statement is a statement that can be written in the form if and only if. If p and q are two statements, the biconditional statement p if and only if q is denoted symbolically by p ↔ q.

Examples

Write the following biconditional in symbolic form.

1. It will rain if and only it is cloudy.

    Let p and q represent the statements:

6   Statements and Connectives

p: It will rain.

q: It is cloudy.

In the symbolic form, the statement is $p \leftrightarrow q$.

2. Afua is not cooking if and only if she is sleeping.

Let p and q represent the statements:

p: Afua is cooking

q: Afua is sleeping.

In the symbolic form, the statement is $\sim p \leftrightarrow q$.

Dominance of Connectives

In symbolic logic statement, grouping is indicated by parentheses. In written logic statement, commas are used to indicate which simple statements are to be grouped together. When we write the compound statement symbolically, the simple statements on the same side of a comma are to be grouped together within parentheses. When no grouping is indicated by parentheses in a symbolic statement or by a comma in written statement, we use the dominance of connectives shown in Table 2.

Table 2

Dominance of connectives
1. Negation, $\sim$
2. Conjunction, $\wedge$ disjunction, $\vee$
3. Conditional, $\rightarrow$
4. Biconditional, $\leftrightarrow$

The table lists the level of dominance of connectives in order from the least dominant connective to the most dominant connective.

When evaluating a symbolic statement that does not contain parentheses, we evaluate the least dominant connective first and the most dominant connective last. The conjunction and disjunction have the same level of dominance. Thus, to indicate which one of these two connectives should be evaluated first, we use parentheses.

For example, the parentheses in the symbolic statement $(p \lor q) \land r$ indicate that disjunction $\lor$ must be evaluated first.

Examples

Write the statements in symbolic form.

1. I will not study, or I will play games and study.

   Let p and q represent the statements:

   p: I will study.

   q: I will play games.

   The placement of the comma indicates that the statement "I will play and study" are to be grouped together. The statement written symbolically with parentheses is $p \lor (q \land p)$.

   The statement is a disjunction because the disjunction symbol is outside the parentheses.

2. If she studies hard then she will attend college or she will start a business.

   Let p, q and r represent the statements:

   p: She studies hard.

   q: She will attend college.

   r: She will start a business.

   Since no commas appear in the sentence, we will evaluate it by using the dominance of connectives. The disjunction has lower dominance, so we place parentheses around the statement "She will attend college or she will start a business". The statement written symbolically with parentheses is $p \rightarrow (q \lor r)$.

3. Add parentheses to the statement $p \lor q \leftrightarrow q \rightarrow \sim p$.

   The biconditional has the greatest dominance, so the parentheses are placed as follows.

   $$(p \lor q) \leftrightarrow (q \rightarrow \sim p)$$

8   Statements and Connectives

This is a biconditional statement because the biconditional symbol is outside the parentheses.

Quantifiers

A word or a phrase that conveys the idea of quantity is called a quantifier. Such words include; all, none (or no), some and not all. The quantifiers: all, every, and each are interchangeable. The quantifiers; some, there exists(s) and at least one are also interchangeable.

Care must be taken when negating statements containing quantifiers. Consider the statement "All birds can fly". This statement is false since at least one bird, the chicken cannot fly. The negation of the statement must be true. The negation of the statement "All birds can fly" is not "No birds can fly" because this statement is also false. The correct negation is "Not all birds can fly" or "At least one bird cannot fly" or "Some birds cannot fly".

Table 3 shows statements that contain quantifiers and their negation.

Table 3

1.All students study maths	Some students do not study maths
2.Some adult own cars	No adult own cars
3. Not all boys can read	All boys can read
4.No students are cleaver	Some students are cleaver

Exercise 1

1. Indicate whether each of the following sentences is a statement or not.

   a) She went to the market.

   b) All the student in the class did the punishment.

   c) Pick the bowl up.

   d) Has he written the letter?

   e) Esi plays football

   f) Is Kwesi studying?

g) Kofi got his passport today.

h) Will you visit your friend?

2. Classify each of the following statement as simple or compound. Write the connective of each compound statement in symbol.

a) Not all students are intelligent.

b) The dog barked or the man walked away.

c) Afua will study if and only if the bulb is fixed.

d) Kofi slipped this afternoon.

e) If it is warm, then I will go swimming.

f) She was the guest speaker.

g) I went home and she went to the super market.

h) She wrote a hit song.

3. Indicate whether each of the following compound statement is a negation, conjunction, disjunction, conditional or biconditional.

a) The colour of the car is blue and the colour of the door is red.

b) The boy is a student or the man works at the factory.

c) The girl is not intelligent.

d) The girl will read a book if and only if she stays at home.

e) If the car breaks down, he will call the mechanics.

f) Adjoa is playing football or Kofi is reading.

g) He will answer the phone if and only if the phone rings.

h) Not everybody likes him.

4. Write the negation of the following statements.

a) Some men have no cars.

b) No shop opens on Sundays.

   c) All babies walk after two years.

   d) Some dogs do not bark

   e) No one wants to buy my phone.

   f) Some people who work hard do not own a house.

5. Let p and q represent respectively the two statements connected by the connectives; and, or, if...then and if and only if. Write each statement in symbolic form.

   a) If he comes home, then his sister will go out.

   b) She will receive good grades or she will go to college.

   c) Esi went to the market and Kwesi worked in the garden.

   d) If it rains today, I will stay indoors.

   e) The concert will be cancelled if and only if it rains.

   f) Esi will go to the movie or Afua will attend the party.

   g) He will be admitted if and only if he gets good grades.

   h) If Adjoa is 12 years old, then she was born on a Sunday.

6. Let p denotes the statement "Ofori went home" and let q denotes the statement "Afua studied French". Write each symbolic statement in words.

   a) $p \lor q$ b) $p \land q$

   c) $\sim p$ d) $p \rightarrow q$

   e) $p \leftrightarrow q$

7. Let p and q represent the statements:

   p: He works hard

   q: He has two cars.

Write each of the following symbolic statements in words.

   a) $\sim p \lor \sim q$ b) $\sim p \rightarrow \sim q$

c) $\sim q \leftrightarrow \sim p$ d) $p \vee \sim q$

e) $\sim p \wedge q$

8. Use the dominance of connectives to add parentheses to each statement. Then indicate whether each statement is a negation, conjunction, disjunction, conditional or biconditional.

     a) $p \wedge \sim q \leftrightarrow \sim p$

     b) $p \vee q \rightarrow r$

     c) $p \leftrightarrow q \rightarrow \sim r$

     d) $q \rightarrow \sim r \vee p$

     e) $p \wedge \sim q \rightarrow \sim p \wedge q$

     f) $p \rightarrow q \vee \sim r$

9. Represent the simple statements by the letters p, q and r. Write each statement symbolical form. Indicate whether the statement is a negation, conjunction, disjunction, conditional or biconditional.

     a) Afua can swim, and if Kofi cannot swim then he can drive.

     b) He will stay home if and only if it rains, or he will go to school.

     c) If the room is hot or the music is loud, then I will stay outside.

     d) Kojo will visit his aunt or he will go to the library, if and only if today is a holiday.

     e) She will have dinner, and she will go to the cinema or she will visit her friend.

10. Represent the simple statements by the letters p, q and r. Write each statement in symbolic form. Place parentheses and, indicate whether the statement is a negation, conjunction, disjunction, conditional or biconditional.

     a) If the moon is out then she is in the garden or she is jogging.

     b) If he stays at home then it is raining or his wife will travel.

c) If she is cleaver and she studies hard then she will get an A.

d) He will pass this course if and only if he studies hard or do his assignment.

e) If she cooks then I will buy lunch if and only if she does not cook early.

2 Truth Tables

The truth value of a statement is either true, denoted by T or false, denoted by F.

A table showing the truth value of a compound statement for all possible cases is called a truth table.

Negation

A statement and its negation have opposite truth values. That is, if p is true, then ~ p is false; if p is false, then ~ p is true. The truth values of ~ p are given in Table 1.

Table 1

p	q
T	F
F	T

Conjunction

Consider the conjunction of any two statements p and q. There are four possible cases to be considered, as illustrated in Table 2. These four cases are listed in the first two columns of the table. The truth values of p $\wedge$ q are given in the third column.

Table 2

p	q	p $\wedge$ q
T	T	T
T	F	F
F	T	F
F	F	F

The conjunction p $\wedge$ q is true whenever both p and q are true statements, and is false in the three remaining cases.

Disjunction

The truth values of p $\vee$ q are given in Table 3.

14   Truth Tables

Table 3

p	q	p ∨ q
T	T	T
T	F	T
F	T	T
F	F	F

The disjunction p ∨ q is false whenever p is false and q is false and that p ∨ q is true otherwise.

Conditional

The truth table for the conditional p → q is given in Table 4.

Table 4

p	q	p → q
T	T	T
T	F	F
F	T	T
F	F	T

Notice that a conditional statement is true in all cases excepts where p is true and q is false.

Biconditional

The truth table for the biconditional p ↔ q is given in Table 5.

Table 5

p	q	p ↔ q
T	T	T
T	F	F
F	T	F
F	F	T

Notice that the biconditional statement is true in the cases where both p and q are true, and where p and q are both false.

Derived Truth Tables

By using the five basic truth tables discussed above, we can construct other truth tables.

Examples

Construct a truth table for each statement.

1. $\sim(\sim p \vee q)$

We construct a table as shown.

The first two columns list all the possible cases. The truth values in the third column are obtained by negating the truth values of p. Next, we use the truth values in the third and second columns and the disjunction table, Table 3 to obtain the truth values in the fourth column. Finally, negating the truth values in the fourth column gives the truth values in the fifth column.

p	q	$\sim p$	$\sim p \vee q$	$\sim(\sim p \vee q)$
T	T	F	T	F
T	F	F	F	T
F	T	T	T	F
F	F	T	T	F

2. $(p \rightarrow q) \wedge \sim q$

p	q	$p \rightarrow q$	$\sim q$	$(p \rightarrow q) \wedge \sim q$
T	T	T	F	F
T	F	F	T	F
F	T	T	F	F
F	F	T	T	T

First list the four cases in the first two columns. In the third column copy the truth values from Table 4. Under $\sim q$, write the negation of the second column. Finally, use the third column and the fourth column and the conjunction table, Table 2 to find the truth values in the fifth column.

16   Truth Tables

Exercise 2

Construct a truth table for each compound statement

1. $p \lor \sim q$

2. $\sim p \lor \sim q$

3. $\sim p \land \sim q$

4. $\sim p \land q$

5. $(p \lor q) \to p$

6. $(p \lor \sim q) \land \sim p$

7. $\sim(\sim p \lor \sim q)$

8. $(p \to q) \leftrightarrow (q \to p)$

9. $(p \lor \sim q) \land (q \land \sim p)$

10. $(\sim p \lor q) \leftrightarrow \sim(p \land \sim q)$

11. $(\sim p \to q) \leftrightarrow (p \lor q)$

12. $(p \land \sim q) \lor (q \land \sim p)$

3 Converse, Inverse and Contrapositive

The converse, inverse and contrapositive are associated to the conditional $p \rightarrow q$.

Converse

Given a conditional statement, you can form another conditional by interchanging the hypothesis and conclusion of the given conditional. This new statement is called the converse of the original conditional. For example, the converse of $p \rightarrow q$ is $q \rightarrow q$.

Table 1 shows the truth values of the conditional statement $p \rightarrow q$ and its converse $q \rightarrow p$.

Table 1

p	q	$p \rightarrow q$	$q \rightarrow p$
T	T	T	T
T	F	F	T
F	T	T	F
F	F	T	T

Notice that in the third row the conditional is true but its converse is false. The converse of a true conditional is not necessarily true.

Example

Consider the statement.

If he attends college, then he is studying mathematics.

Let p and q represent the statements:

p: He attends college.

q: He is studying mathematics.

The statement in symbolic form is $p \rightarrow q$. The converse of this conditional, namely $q \rightarrow p$, states that "if he is studying mathematics, then he attends college."

If the conditional is true, its inverse is not necessarily true. Because he is studying mathematics does not mean that he attends college.

18   Converse, Inverse and Contrapositive

Inverse

The conditional $\sim p \rightarrow \sim q$ is called the inverse of the conditional $p \rightarrow q$.

The truth values for $p \rightarrow q$ and $\sim p \rightarrow \sim q$ are given in Table 2.

Table 2

p	q	$p \rightarrow q$	$\sim p \rightarrow \sim q$
T	T	T	T
T	F	F	T
F	T	T	F
F	F	T	T

Notice that in the third row, the conditional is true while the inverse is false. It follows that the inverse of a conditional is not necessarily true.

If a conditional is true, we cannot conclude that its inverse is true. The inverse of a true conditional is not necessarily true.

Example

Consider the statement

If it rains, then I will not go to the library.

Let p and q represent the statements:

p: It rains.

q: I will not go to the library.

The statement in symbolic form is $p \rightarrow q$. The inverse of this conditional, namely $\sim p \rightarrow \sim q$, states that if it does not rain, then I will go to the library. Because it does not rain does not necessarily mean that I will go to the library.

Contrapositive

The conditional $\sim q \rightarrow \sim p$ is called the contrapositive of the conditional $p \rightarrow q$. The truth values for $p \rightarrow q$ and $\sim q \rightarrow \sim p$ are given in Table 3.

Table 3

p	q	~p	~q	p → q	~q → ~p
T	T	F	T	T	T
T	F	F	T	F	F
F	T	T	F	T	T
F	F	T	T	T	T

From the truth table you will notice that the contrapositive of a true conditional is always true.

Example

Consider the statement

If today is Sunday, then I will watch the game.

Let p and q represents the statements:

p: Today is Sunday.

q: I will watch the game.

The statement in symbolic form is p → q. The contrapositive of this conditional, namely ~ q → ~ p states that if I do not watch the game, then today is not Sunday.

Exercise 3

1. Write the converse, inverse and contrapositive of the conditional statements.

    a) If the light is on, then she will be at home.

    b) If the house is new, then I will buy it.

    c) If the light goes off, then we will go for a walk.

    d) If the pay is not good, then I will not take the job.

    e) If today is a holiday, then the library is not open.

20   Converse, Inverse and Contrapositive

2. Write the converse, inverse and contrapositive of the statements. State whether or not the converse, inverse and contrapositive is true.

   a) If the triangle is isosceles, then two angles are equal.

   b) If two lines intersect in at least one point, then the two lines are not parallel.

   c) If the quadrilateral is a rectangle, then the quadrilateral is a parallelogram.

   d) If the sum of the interior angles of a polygon is $360°$, then the polygon is a quadrilateral.

   e) If the quadrilateral is a square, then the quadrilateral is a rectangle.

   f) If 2 divides a natural number, then 2 divides the unit digit of the number.

3. Write the converse, inverse and contrapositive of each statement:

   a) $\sim p \rightarrow q$

   b) $\sim p \rightarrow \sim q$

   c) $p \rightarrow \sim q$

   d) $(p \wedge q) \rightarrow r$

   e) $(p \wedge \sim q) \rightarrow \sim r$

   f) $(p \vee q) \rightarrow (\sim q \vee r)$

4 Logical Equivalent and Tautology

Equivalent Statements

Statements with the same truth values are said to be logically equivalent (or equivalent).

You can determine whether two statements are equivalent by constructing a truth table for each statement and comparing their truth values. If the statements have identical truth values then the statements are equivalent. If the statements do not have identical truth values then the statements are not equivalent.

Examples

Determine whether the pairs of statements are equivalent.

1. $p \lor q$ and $\sim(\sim p \land \sim q)$

Construct a truth table as shown.

p	q	~p	~q	p∨q	~p∧~q	~(~p∧~q)
T	T	F	F	T	F	T
T	F	F	T	T	F	T
F	T	T	F	T	F	T
F	F	T	T	F	T	F

Because the statements $p \lor q$ and $\sim(\sim p \land \sim q)$ have the same truth values the statements are equivalent.

2. $\sim(p \land q)$ and $\sim p \land \sim q$

Construct a truth table as shown.

p	q	~p	~q	p∧q	~(p∧q)	~p∧~q
T	T	F	F	T	F	F
T	F	F	T	F	T	F
F	T	T	F	F	T	F
F	F	T	T	F	T	T

Because the statements $\sim(p \land q)$ and $\sim p \land \sim q$ do not have the same truth values, the statements are not equivalent.

22   Logical Equivalent and Tautology

Other examples of logically equivalent statements are:

$$\sim(\sim p) \qquad \text{and} \qquad p$$

$$p \to q \qquad \text{and} \qquad \sim q \to \sim p$$

$$q \to p \qquad \text{and} \qquad \sim p \to \sim q$$

$$\sim(p \wedge q) \qquad \text{and} \qquad (\sim p \vee \sim q)$$

$$\sim(p \vee q) \qquad \text{and} \qquad \sim(\sim p \wedge \sim q)$$

$$\sim(p \to q) \qquad \text{and} \qquad (p \wedge \sim q)$$

$$[(p \wedge q) \to r] \qquad \text{and} \qquad [p \to (q \to r)]$$

Tautology

There are some statement patterns that have no false instances, as illustrated in Table 1.

Table 1

p	$\sim p$	$p \vee \sim p$
T	F	T
F	T	T

A compound statement that is true in every possible case is said to be a tautology.

A conditional statement that is a tautology is called an implication. In any implication, if the hypothesis is true, then the conclusion must also be true. That is, if the implication $p \to q$ is true, and if p is true the q must be true.

There are statements whose truth values are all false for each logically possibilities. Such statements are called contradiction.

Examples

Determine whether each statement is a tautology.

1. $p \rightarrow (p \vee q)$

Construct a truth table as shown.

P	q	p ∨ q	p → (p ∨ q)
T	T	T	T
T	F	T	T
F	T	T	T
F	F	F	T

Because the truth values of the statement $p \rightarrow (p \vee q)$ are true in every case the statement is a tautology.

2. $(p \leftrightarrow q) \rightarrow p$

Construct a truth table as shown.

p	q	p ↔ q	(p ↔ q) → p
T	T	T	T
T	F	F	T
F	T	F	T
F	F	T	F

Because the truth values of the statement $(p \leftrightarrow q) \rightarrow p$ are not true in every case the statement is not a tautology.

3. $[(p \wedge q) \wedge \sim q] \rightarrow p$

Construct a truth table as shown.

p	q	p ∧ q	~ q	(p ∧ q) ∧ ~ q	[(p ∧ q) ∧ ~ q] → p
T	T	T	F	F	T
T	F	F	T	F	T
F	T	F	F	F	T
F	F	F	T	F	T

Because the truth values of the statement $p \rightarrow (p \vee q)$ are true in every case the statement is a tautology.

Exercise 4

1. Determine whether the pair of statements are equivalent.

    a) $p \wedge q$, $\sim(\sim p \vee \sim q)$ b) $(p \rightarrow q) \wedge p$, q

    c) $p \rightarrow q$, $\sim(p \wedge \sim q)$ d) $p \leftrightarrow q$, $(p \rightarrow q) \wedge (q \rightarrow p)$

    e) $(p \rightarrow q) \wedge \sim q$, $\sim p$ f) $p \rightarrow q$, $(p \wedge \sim q) \rightarrow (q \wedge \sim q)$

2. Determine whether the statements are tautology.

    a) $[p \wedge (p \vee q)] \leftrightarrow p$ b) $[p \vee (p \wedge q)] \leftrightarrow p$

    c) $[(p \vee q) \wedge \sim p] \leftrightarrow q$ d) $[(p \rightarrow q) \wedge p] \leftrightarrow q$

    e) $(p \rightarrow q) \leftrightarrow (\sim q \rightarrow \sim p)$ f) $\sim(p \rightarrow q) \leftrightarrow (p \wedge \sim q)$

5 Symbolic Arguments

A person presents an argument when he makes a sequence of statements, called hypotheses (or premises) and then draws a conclusion. If the conclusion follows from the hypotheses, we say that our argument is valid. If the conclusion does not follow from the conclusion, we say that our argument is not valid. We can also say that the argument is invalid or the argument is a fallacy.

It is important to note that, when an argument is valid the conclusion necessarily follows from the hypotheses. It is not necessary for the hypothesis or the conclusion to be true statements in an argument.

We generally, write arguments in symbolic form to determine its validity. An argument written in symbolic form is called a symbolic argument.

Valid Arguments

Suppose p and q are the hypotheses of an argument and r the conclusion. We can represent the argument by the conditional statement $(p \wedge q) \rightarrow r$. We can determine the validity of the argument by constructing a truth table for the conditional statement. If the truth value of the statement $(p \wedge q) \rightarrow r$ is true in every case, then the statement is a tautology, and the argument is valid. If the truth value of the statement is not true in every case, then the statement is not a tautology, and the argument is not valid.

Examples

Determine whether the following argument is valid or not valid.

1. If she gets good grades, she will study engineering. She gets good grades. Therefore, she will study engineering.

   First, we separate the hypotheses from the conclusion.

   Hypotheses: If she gets good grades, she will study engineering.

   She gets good grades.

   Conclusion: She will study engineering.

   Next, we write the argument in symbolic form.

   Let p and q represent the statements:

p: She gets good grades.

q: She will study engineering.

Symbolically, the argument is written.

Hypotheses: $p \rightarrow q$

$$p$$

Conclusion: q

Then, we write the argument in the form as shown.

$$[(p \rightarrow q) \wedge p] \rightarrow q$$

Finally, we construct a truth table for this statement.

p	q	$p \rightarrow q$	$(p \rightarrow q) \wedge p$	$[(p \rightarrow q) \wedge p] \rightarrow q$
T	T	T	T	T
T	F	F	F	T
F	T	T	F	T
F	F	T	F	T

Since the truth value of the statement is true in every case, the statement is a tautology. Therefore, the conclusion necessarily follows from the hypotheses and the argument is valid.

2. If today is Tuesday, then I will have lunch. I will not have lunch. Therefore, Today is not Tuesday.

First, we separate the hypotheses from the conclusion.

Hypotheses: If today is Tuesday, then I will have lunch.

I will not have lunch.

Conclusion: Today is not Tuesday.

Let p and q represent the statements:

p: Today is Tuesday.

q: I will have lunch.

Hypotheses: $p \rightarrow q$

$\sim q$

Conclusion: $\sim p$

The argument in symbolic form is

$$[\{p \rightarrow q) \wedge \sim q] \rightarrow \sim p$$

Now, we construct a truth table.

p	q	$\sim p$	$\sim q$	$p \rightarrow q$	$(p \rightarrow q) \wedge \sim q$	$[(p \rightarrow q) \wedge \sim q] \rightarrow \sim p$
T	T	F	F	T	F	T
T	F	F	T	F	F	T
F	T	T	F	T	F	T
F	F	T	T	T	T	T

Since the statement is a tautology, the conclusion necessarily follows from the hypotheses, and argument is valid.

3. She will walk home or she will take a taxi. She did not walk home. Therefore, she takes a taxi.

   Let p and q represent the statements:

   p: She will walk home.

   q: She will take a taxi.

   Hypotheses: $p \vee q$

   $\sim p$

   Conclusion: q

   Construct a truth table as shown.

p	q	$\sim p$	$p \vee q$	$(p \vee q) \wedge \sim p$	$[(p \vee q) \wedge \sim p] \rightarrow q$
T	T	F	T	F	T
T	F	F	T	F	T
F	T	T	T	T	F
F	F	T	F	F	T

Since the statement is not a tautology, the argument is not valid.

28   Symbolic Arguments

Valid Argument Forms

The form of an argument determines its validity. In the preceding examples, we have shown by the use of a true table that an argument in the form

$$[(p \to q) \land p] \to q \qquad \text{or} \qquad [(p \to q) \land \sim q] \to \sim p$$

is valid.

The form of an argument determines its validity. Once we verify that an argument in a particular form is valid, all arguments with exactly the same form will also be valid.

Other valid argument forms are:

1. $(p \to q) \to (\sim q \to \sim p)$

2. $[(p \lor q) \land \sim q] \to p$

3. $[(p \lor q) \land \sim p] \to q$

4. $[p \land (p \to q)] \to q$

5. $[(p \to q) \land (q \to r)] \to (p \to r)$

These argument forms are called rules of inference (or laws of logic).

You can use a valid argument form to draw logical conclusion.

Example

Determine a logical conclusion that follows from the given statements.

If you order the book online, you will get it at 2 p.m.

You did not get the book at 2 p.m.

If we consider the argument to be of the form

$$[(p \to q) \land \sim q] \to \sim p,$$

then the logical conclusion is °Therefore, you did not order the book online."

Direct Proof

In any valid proof one shows that the implication $p \rightarrow q$ is a tautology. To do this by the method of direct proof, we assume that p is true and show that q must be true.

Examples

Determine whether each argument is valid or not valid.

1. Hypotheses: $\sim p \rightarrow \sim q$

$$p$$

   Conclusion: q

   We assume $\sim p \rightarrow \sim q$ and p are both true. Since $\sim p \rightarrow \sim q$ is true, its contrapositive, $q \rightarrow p$ is also true. Since p is true, q must be true. Therefore, the argument is valid.

2. Hypotheses: $p \vee q$

$$\sim p$$

   Conclusion: q

   We assume $p \vee q$ and $\sim p$ are both true. Since $\sim p$ is true, then p is false. Since $p \vee q$ is true and p is false, q must be true. Therefore, the argument is valid.

3. If I study hard, then I will make good grades.

   I did not make good grades. Therefore, I did not study hard.

   Let p and q represent the statements:

   p: I study hard.

   q: I will make good grades.

   In symbolic form, the argument is

   Hypotheses: $p \rightarrow q$

$$\sim q$$

Conclusion: ~p

We assume $p \rightarrow q$ and ~q are both true. Since ~q is true, q is false. Since $p \rightarrow q$ is true and q is false, p is false. Since p is false, ~p is true, so the argument is valid.

4.   He will study or he will wash the car. He did not wash the car. Therefore, he did not study.

Let p and q represent the statements:

P: He will study.

q: He will wash the car.

Hypotheses: $p \lor q$

$\sim q$

Conclusion: ~p

We assume $p \lor q$ and ~q are both true. Since ~q is true, q is false. Since $p \lor q$ is true and q is false, p must be true. Since p is true, ~p is false, so the argument is not valid.

Indirect Proof

To determine the validity of the argument $p \rightarrow q$ by the method of indirect proof, we assume that q is false and show that this will lead to a contradiction, that is, logically impossible situation. We resolve the contradiction by only abandoning the assumption that the conclusion is false. That is, the conclusion must be true. An indirect proof is also called a proof by contradiction.

Examples

Prove that the conclusion of each argument is true.

1. Hypotheses: $p \lor q$

$\sim q$

Conclusion: p

We assume that $p \lor q$ and ~q are true.

We went to show that the conclusion p is true. We assume that p is false, that is ~ p is true, and then shows that this leads to contradiction. Since p ∨ q is true, either p is true or q is true, or both p and q are true. But both p and q are false, which is impossible. The assumption p is false leads to a contradiction. Thus, p is true.

2. If I walk home, I will pick Ama from school. I did not pick Ama from school. Therefore, I did not walk home.

Let p and q represent the statements:

p: I walk home.

q: I will pick Ama from school.

In symbolic form the argument is

Hypotheses: p → q

                 ~ q

Conclusion: ~ p

We want to show that the conclusion ~ p is true. We suppose that ~ p is false. Thus, we assume that p → q, ~ q and p are all true. Since p → q is true, its contraposition ~ q → ~ p is true. If ~ q and ~ q → ~ p are true, then ~ p is true. But ~ p is both true and false, which is impossible. The assumption ~ p is false leads to a contradiction. Thus, ~ p must be true.

Exercise 5

1. Use truth table analysis to prove that each of the following statements is logically valid.

    a) (p ∨ q) ↔ ~ (~ p ∧ ~ q)

    b) (p → q) ↔ ~ (p ∧ ~ q)

    c) [(p → q) ∧ ~ q] → q

    d) [(p ∧ q) ∨ ~ q] → p

    e) [p ∧ (p → q)] → q

32   Symbolic Arguments

f) $(p \leftrightarrow q) \leftrightarrow [(p \rightarrow q) \wedge (q \rightarrow p)]$

2. Use a truth table to determine whether each argument is valid or not valid.

a) Hypotheses: I will play outside or I will go for a walk.

                       I did not play outside.

  Conclusion: I went for a walk

b) Hypotheses: If he gets good grades, he will study science.

                       He studied science.

  Conclusion: He did not get good grades.

c) Hypotheses: I will win the race if and only if I train regularly

                       I trained regularly.

  Conclusion: I won the race.

d) Hypotheses: It is not cloudy or it is not raining.

                       It is cloudy.

  Conclusion: It is not raining.

3. Determine whether the argument is valid or not valid, by direct proof.

a) Hypotheses: If it rains, then I will stay at home

                       I did not stay at home.

  Conclusion: It did not rain.

b) Hypotheses: If I practise a lot, I will pass the maths exam.

                       I did not practise a lot.

  Conclusion: I did not pass the maths exam.

c) Hypotheses: I will study hard or I will fail the course.

   I did not fail the course.

   Conclusion: I studied hard.

d) Hypotheses: If it is sunny, I will play outside.

   I did not play outside.

   Conclusion: It was not sunny.

4. Write each argument in symbolic form. Then determine whether the argument is valid or not valid.

   a) If it rains, she will not go to the market. She went to the market. Therefore, it did not rain.

   b) If I do not go to school, I will go shopping. I did not go shopping. Therefore, I went to school.

   c) Kwesi will read or he will go for a walk. Kwesi did not go for a walk. Therefore, he read.

   d) The test was not easy or I did not receive a good grade. I received a good grade. Therefore, the test was easy.

5. Determine whether the conclusion of each argument is true, by the use of the proof contradiction.

   a) If I rest well, then I can work hard. I did nor rest well. Therefore I did not work hard.

   b) If today is Tuesday, then I will travel. I did not travel. Therefore, today is not Tuesday.

   c) Kofi can drive or he owns a car. Kofi does not own a car. Therefore, Kofi can drive.

   d) You will cook lunch. If you cook lunch, then I will visit my mother. Therefore, I visit my mother.

6. Using a valid argument form, supply a logical conclusion to the following arguments. Verify that the argument is valid for the conclusion you supplied.

34   Symbolic Arguments

a)   If you study hard, then you will get good grades. You study hard.

b)   If you order the computer online, you must pay with debit card. You did not pay with a debit card.

c)   If he does not go swimming, then it is not warm. It is warm.

d)   I will pay my tuition or I will get a scholarship. I did not get a scholarship.

e)   If you sell the house, then you will get a commission. You did not get a commission.

f)   You did not receive your salary or you did not buy groceries. You bought groceries.

6 Deductive Reasoning

The laws of logic and relevant definitions can be used to provide a system for reaching logical conclusions, called deductive reasoning. Deductive reasoning is a process of drawing a conclusion based on hypotheses that are generally assumed to be true. In logic, these hypotheses are definitions and laws of logic.

Listed below are some laws of logic with their names.

1. $\sim(\sim p) \equiv p$ Double Negation

2. $p \rightarrow q \equiv \sim p \vee q$ Definition 1

3. $(p \leftrightarrow q) \equiv (p \rightarrow q) \wedge (q \rightarrow p)$ Definition 2

4. $p \rightarrow q \equiv \sim q \rightarrow \sim p$ Contrapositive law

5. $p \wedge q \equiv q \wedge p$ $p \vee q \equiv q \vee p$ Commutative laws

6. $\sim(p \wedge q) \equiv \sim p \vee \sim q$ De Morgan's laws

    $\sim(p \vee q) \equiv \sim p \wedge \sim q$

7. $(p \wedge q) \wedge r \equiv p \wedge (q \wedge r)$ Associative laws

    $(p \vee q) \vee r \equiv p \vee (q \vee r)$

Examples

1. Determine a statement equivalent to:

    a) $\sim(\sim p \rightarrow q)$

        $p \rightarrow q \equiv (\sim p) \vee q$ Definition 1

        $\sim(p \rightarrow q) \equiv \sim[(\sim p) \vee q]$ Negate both side

                $\equiv p \wedge \sim q$ De Morgan's law

       Hence $\sim(\sim p \rightarrow q) \equiv p \wedge \sim q$

b) $p \wedge q$

$$p \wedge q \equiv\, \sim (\sim p) \,\wedge \sim (\sim q) \qquad \text{Double Negation}$$

$$\equiv\, \sim (\sim p \,\vee \sim q) \qquad \text{De Morgan's law}$$

$$\equiv\, \sim (p \rightarrow\, \sim q) \qquad \text{Definition 1}$$

2. Write a statement equivalent to:

"If he does not study, then he will go swimming."

Let   p: He studies.

           q: He will go swimming.

The given statement can be written symbolically as $\sim p \rightarrow q$. This statement is equivalent to $p \vee q$. Therefore, an equivalent statement is "He studies or he will go swimming."

3. It is false that if he is a carpenter, then he is not rich.

Let   p: He is a carpenter.

           q: He is rich.

The given statement can be represented symbolically as $\sim (p \rightarrow\, \sim q)$. This statement is equivalent to $p \wedge q$. Therefore, an equivalent statement is "He is a carpenter and he is rich."

4. Determine if a valid conclusion can be reached from the two statements.

"If the number is even, then it is divisible by 2" and "If the number is not even, then it is odd."

Let p, q and r represent the statements:

p: The number is even.

q: The number is divisible by 2.

r: The number is odd.

The given statements can be written symbolically as $p \to q$ and $\sim p \to r$. The statement $p \to q$ is equivalent to $\sim q \to \sim p$. Now, $\sim q \to \sim p$ and $\sim p \to r$, so by the law of syllogism $\sim q \to r$. The statement $\sim q \to r$ is "If the number is not divisible by 2, then the number is odd."

We can use deductive reasoning to construct valid proofs.

Examples

Prove the following by deductive reasoning.

1. $p \to q \equiv \sim q \to \sim p$

$$p \to q \equiv \sim p \vee q \qquad\qquad \text{Definition 1}$$

$$\equiv q \vee \sim p \qquad\qquad \text{Commutative law}$$

$$\equiv \sim (\sim q) \vee \sim p \qquad\qquad \text{Double Negation}$$

$$\equiv \sim q \to \ \sim p \qquad\qquad \text{Definition 1}$$

Hence, $p \to q \equiv \sim q \to \sim p$

2. $p \to (q \vee r) \equiv (p \wedge \sim q) \to r$

$$p \to (q \vee r) \equiv \sim p \vee (q \vee r) \qquad\qquad \text{Definition 1}$$

$$\equiv (\sim p \vee q) \vee r \qquad\qquad \text{Associative law}$$

$$\equiv \sim (p \wedge \sim q) \vee r \qquad\qquad \text{De Morgan's law}$$

$$\equiv (p \wedge \sim q) \to r \qquad\qquad \text{Definition 1}$$

Hence, $p \to (q \vee r) \equiv (p \wedge \sim q) \to r$

3. $p \rightarrow (q \rightarrow r) \equiv (p \wedge q) \rightarrow r$

$$p \rightarrow (q \rightarrow r) \equiv p \rightarrow (\sim q \vee r) \qquad \text{Definition 1}$$

$$\equiv \sim p \vee (\sim p \vee r) \qquad \text{Definition 1}$$

$$\equiv (\sim p \vee \sim q) \vee r \qquad \text{Associative law}$$

$$\equiv \sim (p \wedge q) \vee r \qquad \text{De Morgan's law}$$

$$\equiv (p \wedge q) \rightarrow r \qquad \text{Definition 1}$$

Hence, $p \rightarrow (q \rightarrow r) \equiv (p \wedge q) \rightarrow r$

4. $(p \wedge q) \rightarrow r \equiv (p \wedge \sim r) \rightarrow \sim q$

$$(p \wedge q) \rightarrow r \equiv \sim (p \wedge q) \vee r \qquad \text{Definition 1}$$

$$\equiv \sim p \vee \sim q \vee r \qquad \text{De Morgan's law}$$

$$\equiv \sim p \vee r \vee \sim q \qquad \text{Commutative law}$$

$$\equiv \sim (p \wedge \sim r) \vee \sim q \qquad \text{De Morgan's law}$$

$$\equiv (p \wedge \sim r) \rightarrow \sim q \qquad \text{Definition 1}$$

Hence, $(p \wedge q) \rightarrow r \equiv (p \wedge \sim r) \rightarrow \sim q$

Exercise 6

1. Using the fact that $p \rightarrow q$ is equivalent to $\sim p \vee q$ write an equivalent form of the statement.

   a) If you go out at 12 p.m., then you will miss lunch.

   b) She is not cooking or she is washing the dishes.

   c) If she does not play games often, she will study.

   d) Afua has retired or she is still working.

e) If the air conditioner is on, then the room is warm.

f) Kojo does not receive his salary or he will buy me a gift.

2. Use the contrapositive law to write equivalent statements to:

   a) If he is a college graduate, then he can apply for the job.

   b) If she passes the bar exam, then I will buy her a car.

   c) If the man is not at the gate, then the dog will not bark.

   d) If the bus is overbooked, then I will not travel.

   e) If she is not well known, then she is not a singer.

3. Determine if a valid conclusion can be reached from the two statements.

   a) "If he gets a job at the factory, then he will earn good salary" and "If he earns good salary, then he will buy a car."

   b) "If he does not rehearse well, then he will not enter the competition" and "If he rehearses well, then he will win the top prize."

   c) "If he does not pass the entrance exam, then he will not go to college" and "If he passes the entrance exam, he will go on tour."

   d) "If I do not go to the concert, then I will write an article" and "If I go to the concert, then I will have dinner at the restaurant."

   e) "If I do not get a discount, then I will not buy the shirts" and "If I get a discount, then I will have lunch."

4. Prove the following by deductive reasoning.

   a) $p \vee q \equiv \sim (\sim p \wedge \sim q)$

   b) $\sim (p \wedge q) \equiv \sim p \vee \sim q$

   c) $p \rightarrow q \equiv \sim (p \wedge \sim q)$

   d) $\sim p \rightarrow q \equiv \sim q \rightarrow p$

e) $q \rightarrow p \equiv {\sim} q \vee p$

f) ${\sim}({\sim} q \rightarrow {\sim} p) \equiv p \wedge {\sim} q$

g) $q \rightarrow {\sim}(p \wedge {\sim} r) \equiv q \rightarrow ({\sim} p \vee r)$

h) $({\sim} p \vee {\sim} q) \rightarrow r \equiv {\sim}(p \wedge q) \rightarrow r$

7 Syllogistic Arguments

Syllogistic logic (or syllogism) is deductive process of arriving at a conclusion, using diagrams. In syllogistic logic, we use statements containing the quantifiers: all, some, no and none.

Venn Diagrams

Venn diagrams are used to show relationship between and among sets. Venn diagrams consist of circles enclosed in a rectangle. The rectangle is used to represent the universal set, and the circles are used to represent a subset of the universal set.

Here, we will use Venn diagrams to illustrate relationships described in statements in the form:

All As are Bs

No As are Bs

Some As are Bs

Some As are not Bs

Examples

Draw Venn diagrams to illustrate the following statements.

1. All boys in the class play football.

Let U = {students who are in the class}

A = {boys who are in the class}

B = {students who play football}

The set B contains all elements in set A, so A is a subset of B. A Venn diagram illustrating this relationship is as shown on the next page.

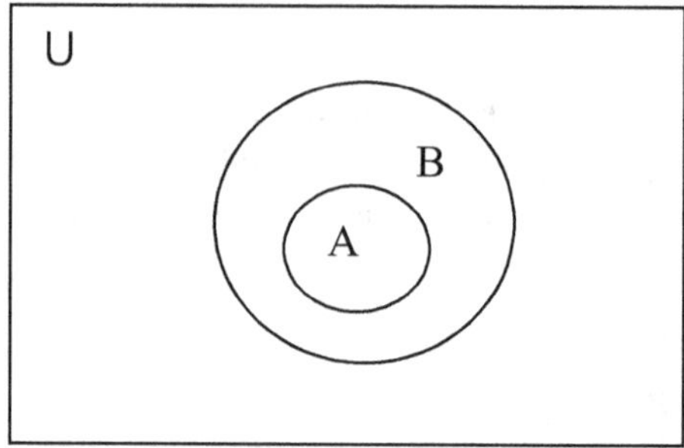

The inner circle represents all the boys in the class and the outer circle represents all students who play football.

2. No rich men take taxis.

Let U = {all people}

    A = {men who are rich}

    B = {people who take taxis}

The set A and set B have no common elements. That is, set A and set B are disjoint sets, and the relationship is illustrated as shown.

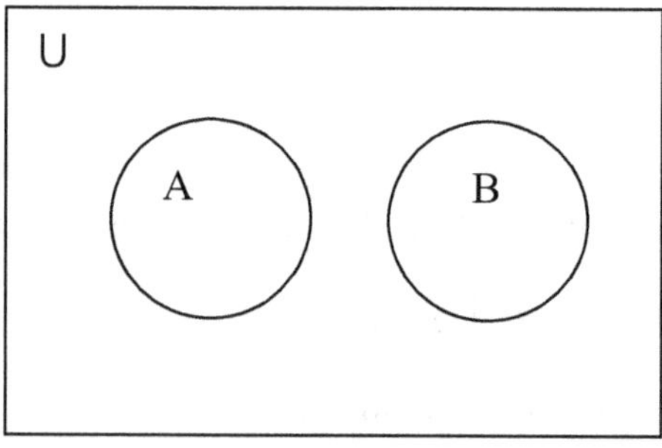

3. Some policemen wear uniform.

Let U = {all people}

    A = {all policemen}

    B = {people who wear uniform}

There is at least one element common to set A and set B. In the Venn diagram, the shaded region represents policemen who wear uniform.

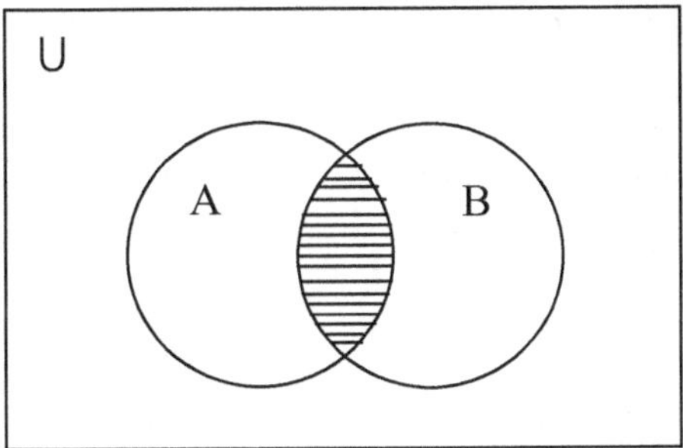

4. Some children do not like ice cream.

Let U = {all people}

A = {all children}

B = {children who like ice cream}

There is at least one element that is in set A that is not in set B. In the Venn diagram, the shaded region represents children who do not like ice cream.

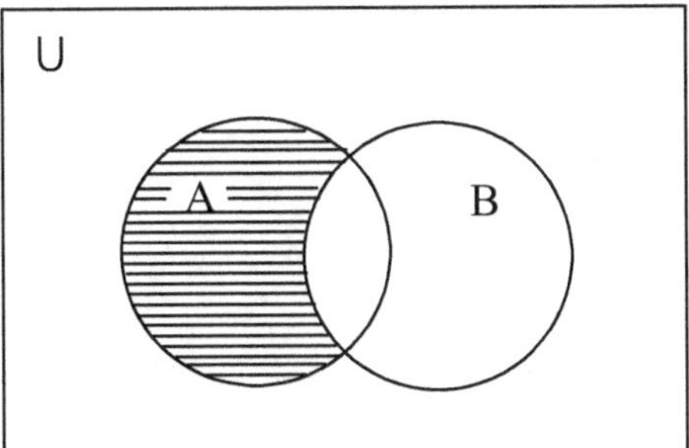

Euler Diagrams

An Euler diagram is similar to a Venn diagram. While Venn diagrams show all possible relationship described in a statement, Euler diagrams show the exact relationship described in a statement.

44   Syllogistic Reasoning

Validity of a Syllogistic Argument

The validity of a syllogistic argument is determined by using Euler diagram. Like symbolic logic, syllogistic logic consists of a set of statements that contain quantifiers, called hypotheses followed by a conclusion. Recall that an argument is valid when its conclusion necessarily follows from a given set of hypotheses. An argument in which the conclusion does not necessarily follow from the given hypotheses is said to be not valid.

Examples

Determine whether the following syllogistic logic is valid or is not valid.

1. Hypotheses: All science students are intelligent.

                          All intelligent students are girls.

    Conclusion: All science students are girls.

Let I = {student who are intelligent}

    S = {students who study science}

    G = {students who are girls}

The Euler diagram illustrating the statements is as shown.

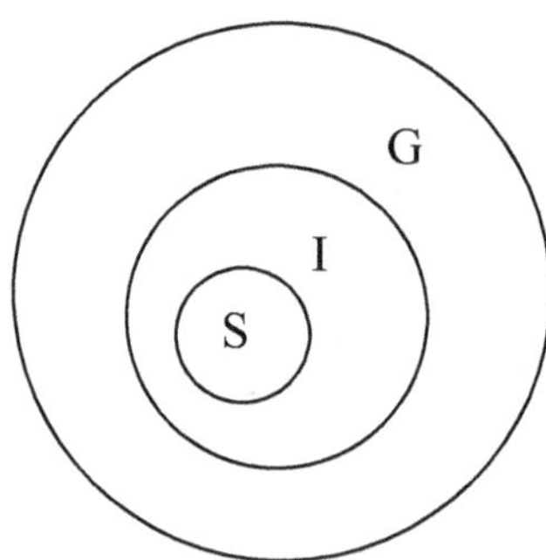

Since the set S is within G, we conclude that all science students are girls. Therefore, the argument is valid, since the conclusion "All science students are girls" necessarily follows from the set of hypotheses.

2. Hypotheses: All girls like reading.

Kojo is a girl.

Conclusion: Kojo likes reading.

Let R = {people who like reading}

G = {girls who like reading}

The set G is a subset of R and the statement "All girls like reading" is illustrated as shown.

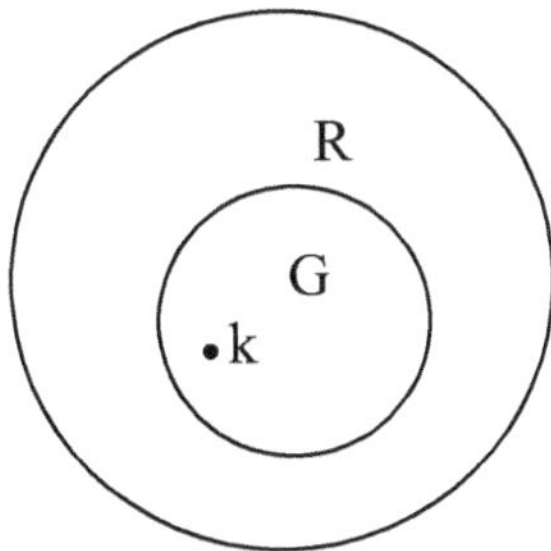

The point labelled k in the inner circle represents the statement "Kojo is a girl." Since the point k is within the set R, we conclude that Kojo likes reading. Therefore, the argument is valid.

3. Hypotheses: All rich men own cars

Afua own cars

Conclusion: Afua is a rich man

Let G = {people who own cars}

R = {men who are rich}

The set R is a subset of the set G and the statement

"All rich men own cars"

is illustrated as shown on the next page.

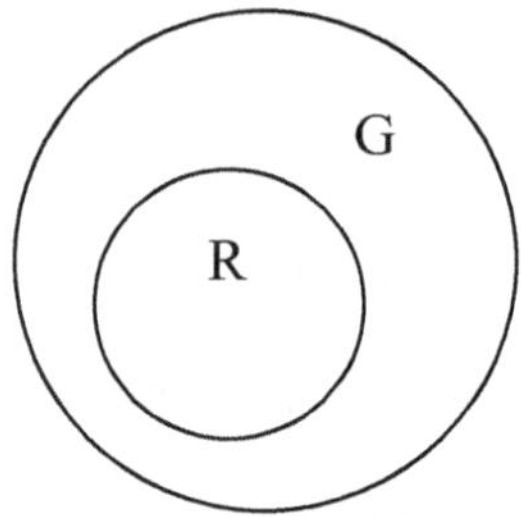

A point which we will label *a*, representing the statement "Afua own cars", can be placed either in the inner circle or the outer circle as shown.

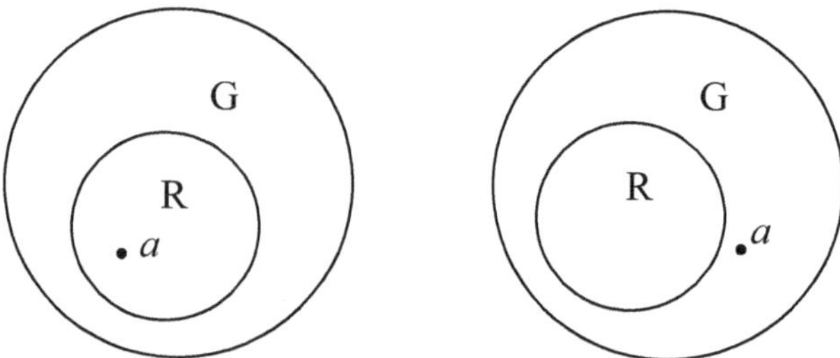

You can see from the second diagram that Afua is not a rich man. Since the conclusion does not necessarily follow from the hypotheses, the argument is not valid.

4. Hypotheses: No lazy students pass this course.

   Intelligent students pass this course.

Conclusion: No lazy students are intelligent.

Let L = {students who are lazy}

   C = {students who pass the course}

   I = {students who are intelligent}

The first statement shows that the set L and the set C are disjoint sets. The second statement shows that the set I is a subset of the set C. The two statements are illustrated in a diagram as shown.

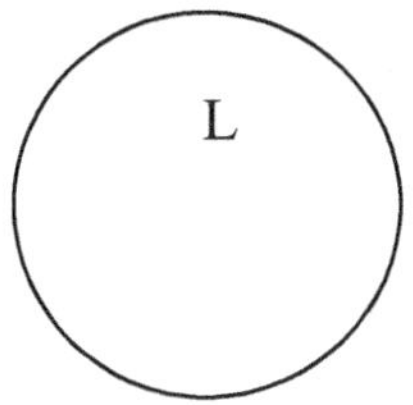 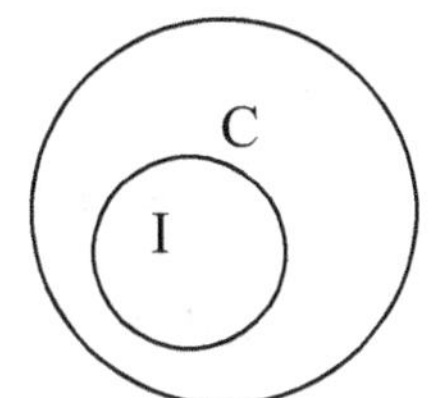

The sets L and I are disjoint. Thus, no lazy student can be intelligent, and the argument is valid.

Exercise 7

Use an Euler diagram to determine whether the syllogistic is valid or not valid.

1. All living things walk.

    An orange tree is a living thing.

    Therefore, an orange tree walks.

2. All birds sing.

    Mercy is a bird.

    Therefore, Mercy sings.

3. All doctors are healthy.

    Ama is healthy.

    Therefore, Ama is a doctor.

4. All girls are bullies.

    Kwesi is a girl.

    Therefore, Kwesi is a bully.

5. Some knowledgeable adults are teachers.

    Mensah is a teacher.

    Therefore, Mensah is knowledgeable.

6. All students are intelligent.

   No intelligent students are lawyers.

   All lawyers are not students.

7. All teachers wear glasses.

   No students wear glasses.

   No teachers are students.

8. All bicycle riders wear helmets.

   Some helmets are expensive.

   Some bicycle riders buy expensive helmets.

Review Exercises

1. Determine which of the following sentences are statements.

   a) Esi will sell her car.

   b) When will he come?

   c) The boy is intelligent.

   d) Akosua comes home every week.

   e) Take the child home.

2. Classify each of the following statements as simple or compound. Name the connective of each compound statement.

   a) Mensah will sell his old television set and buy a new one.

   b) She is beautiful.

   c) The boy is intelligent or he is talented.

   d) If the examination is hard, the grades are low.

   e) The baby is crying.

   f) I will stay home if and only if it is raining.

3. Negate the following statements.

   a) I will study hard.

   b) Esi is talented.

   c) I will wash my car.

   d) Afua will do her homework.

4. Construct truth tables for the statements.

   a) $p \wedge (q \vee \sim q)$

   b) $(p \vee q) \wedge (p \wedge \sim q)$

   c) $[(p \rightarrow q) \wedge \sim q] \rightarrow \sim p$

   d) $(p \wedge q) \vee (\sim p \wedge \sim q)$

5. Put each statement into symbolic form.

   a) I will study or I will go fishing.

   b) The boy is intelligent and he will make a good grade.

   c) I will dance if and only if she sings.

   d) If I study hard, I will pass the test.

6. Write the converse, inverse and contrapositive of the statements.

   a) If it is cloudy, then it is raining.

   b) If I walk home, I will miss the program.

   c) If Esi does not obey school rules, she will be expelled.

   d) If I go to school, I will not go fishing.

7. Use truth tables to determine whether the following arguments are equivalent.

   a) $\sim p \vee q,\ \ p \rightarrow q$ b) $\sim (p \wedge q),\ \ \sim p \wedge q$

   c) $\sim (\sim p \vee q),\ \ p \wedge \sim q$ d) $\sim (p \rightarrow \sim q),\ p \wedge q$

   e) $(p \rightarrow q) \wedge (\sim q \vee p),\ \ p \leftrightarrow q$ f) $(p \rightarrow q) \wedge (q \rightarrow p),\ q \leftrightarrow p$

8. Use truth tables to determine whether the following arguments are valid or not valid.

   a) He is a maths teacher or he is a high school teacher. He is not a maths teacher. Therefore, he is not a high school teacher.

   b) If Kojo is innocent, he will not appear before the disciplinary committee. Kojo appears before the disciplinary committee. Therefore, Kojo is not innocent.

   c) If today is a holiday, Kofi will not go to school. Kofi does not go to school. Therefore, today is a holiday.

   d) If the pay is good, then I will take the job. If I take the job, then I will pay the rent. Therefore, if the pay is good, then I will pay the rent.

e)   If the house has a garden, then I will buy the house. If the house does not have a garden, then I will invest the money in a savings account. Therefore, if I do not buy the house, then I will invest the money in a savings account.

9.   Determine whether the argument is valid, by the use of direct proof.

   a) If I do not work outside, I will paint the hall. I will not paint the hall. Therefore, I will work outside.

   b) Ama will eat breakfast if she gets up early. She gets up early. Therefore, Ama eats breakfast.

   c) It is sunny or it is not cloudy. It is cloudy. Therefore, it is not sunny.

   d) If Kojo passes the final exam, then he will go to college. Kojo do not go to college. Therefore, Kojo do not pass the final exam.

   e) If the shirt is new, then I will wear it tonight. I do not wear the shirt. Therefore, the shirt is not new.

10. Determine whether the conclusion of each argument is true, by the use of proof of contradiction.

   a) If you do your homework regularly, then you get an A. You do your homework regularly. Therefore, you will get an A.

   b) Afua reads a novel or she will go for a walk. Afua does not read a novel. Therefore, she will not go for a walk.

   c) If you pay with a credit card, then you will get a discount. You did not get a discount. Therefore, you did not pay with a credit card.

   d) If you do not practise a lot, then you will not get an A in maths. You practise a lot. Therefore, you will get an A in maths.

   e) If the tuition fee is too high, then I will pay the fee. If the tuition fee is not too high, then I will not defer the course. If I do not pay the fee, then I will not defer the course.

11. Prove the following by deductive reasoning.

a) $\sim (p \lor q) \equiv \sim p \land \sim q$

b) $p \to q \equiv \sim p \lor q$

c) $p \to \sim q \equiv q \to \sim p$

d) $\sim [(p \to q) \lor (q \to p)] \equiv p \leftrightarrow q$

12. Use an Euler diagram to determine whether the syllogistic is valid or not valid.

a) All dogs bark

A cat is a dog.

Therefore, a cat barks.

b) All children like ice cream

My dog likes ice cream.

My dog is a child.

c) No public servant own guns

All policemen own guns.

Therefore, no public servants are policemen

d) All students can read.

Some adults are students.

Therefore, some adults can read.

e) All actors are famous

Some famous people are rich

Some famous actors are rich

Answers to Exercises

Exercise 1

1. a) Statement b) statement c) not d) not
   e) statement f) statement h) not

2. a) compound b) compound c) compound
   d) simple e) compound f) simple
   g) compound h) simple

3. a) conjunction b) disjunction c) negation
   d) biconditional e) conditional f) disjunction
   g) biconditional h) negation

4. a) All men have cars.

   b) Some shop opens on Sunday.

   c) Some babies do not walk after two years.

   d) No dogs bark.

   e) Someone wants to buy my phone.

   f) All people who work hard own a house.

5. a) $p \rightarrow q$ b) $p \vee q$ c) $p \wedge q$ d) $p \rightarrow q$
   e) $p \leftrightarrow q$ f) $p \vee q$ g) $p \leftrightarrow q$ h) $p \rightarrow q$

6. a) Ofori went home or Afua studied French.

   b) Ofori went home and Afua studied French.

   c) Ofori did not go home.

   d) If Ofori went home, then Afua studied French.

   e) Ofori went home if and only if Afua studied French.

7. a) He does not work hard or he does not have two cars.

   b) If he does not work hard, then he does not have two cars.

   c) If he does not have two cars, then he does not work hard.

   d) He works hard or he does not have two cars.

   e) He does not work hard and he has two cars.

8. a) $(p \wedge \sim q) \leftrightarrow \sim p$ b) $(p \vee q) \to r$

   c) $p \leftrightarrow (q \to \sim r)$ d) $q \to (\sim r \vee p)$

   e) $(p \wedge \sim q) \to (\sim p \wedge q)$ f) $p \to (q \vee \sim r)$

9. a) $p \wedge (\sim q \to r)$ conjunction   b) $(p \leftrightarrow q) \vee r$ disjunction

   c) $(p \vee q) \to r$ conditional   d) $(p \vee q) \leftrightarrow r$ biconditional

   e) $p \wedge (q \vee r)$ conjunction

10. a) $p \to (q \vee r)$ conditional   b) $p \to (q \vee r)$ conditional

   c) $(p \wedge q) \to r$ conditional   d) $p \leftrightarrow (q \vee r)$ biconditional

   e) $(p \to q) \leftrightarrow r$ biconditional

Exercise 2

1.

p	q	$\sim q$	$p \vee \sim q$
T	T	F	T
T	F	T	T
F	T	F	F
F	F	T	T

2.

p	q	$\sim p$	$\sim q$	$\sim p \vee \sim q$
T	T	F	F	F
T	F	F	T	T
F	T	T	F	T
F	F	T	T	T

3.

p	q	~p	~q	~p ∧ ~q
T	T	F	F	F
T	F	F	T	F
F	T	T	F	F
F	F	T	T	T

4.

p	q	~p	~p ∧ q
T	T	F	F
T	F	F	F
F	T	T	T
F	F	T	F

5.

p	q	p ∨ q	(p ∨ q) → p
T	T	T	T
T	F	T	T
F	T	T	F
F	F	F	T

6.

p	q	~p	~q	p ∨ ~q	(p ∨ ~q) ∧ ~p
T	T	F	F	T	F
T	F	F	T	T	F
F	T	T	F	F	F
F	F	T	T	T	T

7.

p	q	~p	~q	~p ∨ ~q	~(~p ∨ ~q)
T	T	F	F	F	T
T	F	F	T	T	F
F	T	T	F	T	F
F	F	T	T	T	F

8.

p	q	p → q	q → p	(p → q) ↔ (q → p)
T	T	T	T	T
T	F	F	T	F
F	T	T	F	F
F	F	T	T	T

9.

p	q	~p	~q	P ∨ ~q	Q ∧ ~p	(p ∨ ~q) ∧ (q ∧ ~p)
T	T	F	F	T	T	T
T	F	F	T	T	F	F
F	T	T	F	F	T	F
F	F	T	T	T	T	T

10.

p	q	~p	~q	~p ∨ q	P ∧ ~q	~(p ∧ ~q)	(~p ∨ q) ↔ ~(p ∧ ~q)
T	T	F	F	T	F	T	T
T	F	F	T	F	T	F	T
F	T	T	F	T	F	T	T
F	F	T	T	T	F	T	T

11.

p	q	~p	~p → q	p ∨ q	(~p → q) ↔ (p ∨ q)
T	T	F	F	T	F
T	F	F	T	T	T
F	T	T	T	T	T
F	F	T	T	F	F

12.

p	q	~p	~q	p ∧ ~q	q ∧ ~p	(p ∧ ~q) ∨ (q ∧ ~p)
T	T	F	F	F	F	F
T	F	F	T	T	F	T
F	T	T	F	F	T	T
F	F	T	T	F	F	F

Exercise 3

1. a) Converse: If she is at home, then the light will be on.

   Inverse: If the light is not on, then she will not be at home.

   Contrapositive: If she is not at home, then the light is not on.

   b) Converse: If I buy the house, then the house is new.

   Inverse: If the house is not new, then I will not buy it.

   Contrapositive: If I do not buy the house, then the house is not new

   c) Converse: If I go for a walk, then the light goes off.

   Inverse: If the light does not go off, then I will not go for a walk.

   Contrapositive: If I do not go for a walk, then the light does not go off.

   d) Converse: If I do not take the job, then the pay is not good.

   Inverse: If the pay is good, then I will take the job.

   Contrapositive: If I take the job, then the pay is good.

   e) Converse: If the library is not open, then today is a holiday.

   Inverse: If today is not a holiday, then the library is open.

   Contrapositive: If the library is open, then today is not a holiday.

2. a) Converse: If two angles of a triangle are equal, then the triangle is isosceles. Not

   Inverse: If the triangle is not isosceles, then two angles are not equal. Not

   Contrapositive: If two angles of a triangle are not equal, then the triangle is not isosceles. True

b) Converse: If two lines are not parallel, then the two lines intersect in at least one point.   Not

Inverse: If two lines do not intersect in at least one point, then the two lines are parallel.   Not

Contrapositive: If two lines are parallel, then the two lines do not intersect in at least one point.   True

c) Converse: If the quadrilateral is a parallelogram, then the quadrilateral is a rectangle.   Not

Inverse: If the quadrilateral is not a rectangle, then the quadrilateral is not a parallelogram. Not

Contrapositive: If the quadrilateral is not a parallelogram, then the quadrilateral is not a rectangle. True

d) Converse: If the polygon is a quadrilateral, then the sum of the sum of the interior angles of the polygon is $360°$. True

Inverse: If the sum of the interior angles of a polygon is not $360°$, then the polygon is not a quadrilateral. True

Contrapositive: If the polygon is not a quadrilateral, then the sum of the interior angles of the polygon is not $369°$. True

e) Converse: If the quadrilateral is a rectangle, then the quadrilateral is a square. Not

Inverse: If the quadrilateral is not a square, then the quadrilateral is not a rectangle. Not

Contrapositive: If the quadrilateral is not a rectangle, then the quadrilateral is not a square. True

f) Converse: If 2 divides the natural number, then 2 divides the unit digit is of the number.   True

Inverse: If 2 does not divide the unit digit of the natural number, then 2 does not divide the number.   True

Contrapositive: If 2 does not divide the natural number, then 2
does not divide the unit digit of the number.
True

3.   Converse   Inverse   Contrapositive

a) $q \rightarrow \sim p$ $p \rightarrow \sim q$ $\sim q \rightarrow p$

b) $\sim q \rightarrow \sim p$ $p \rightarrow q$ $q \rightarrow p$

c) $\sim q \rightarrow p$ $\sim p \rightarrow q$ $q \rightarrow \sim p$

d) $r \rightarrow (p \wedge q)$ $\sim (p \wedge q) \rightarrow \sim r$ $r \rightarrow \sim (p \wedge \sim q)$

e) $\sim r \rightarrow (p \wedge \sim q)$ $\sim (p \wedge \sim q) \rightarrow r$ $r \rightarrow \sim (p \wedge \sim q)$

f) Converse: $(\sim q \vee r) \rightarrow (p \vee q)$

Inverse: $\sim (p \vee q) \rightarrow \sim (\sim q \vee r)$

Contrapositive: $\sim (\sim q \vee r) \rightarrow \sim (p \vee q)$

Exercise 4

1. a) Equivalent   b) Not   c) Equivalent

   d) Equivalent   e) Not   f) Equivalent

2. a) Tautology   b) Tautology   c) Not

   d) Not   e) Tautology   f) Tautology

Exercise 5

1. a) valid   b) valid   c) not valid   d) not valid

   e) valid   f) valid

2. a) valid   b) not valid   c) valid   d) valid

3. a) valid   b) not valid   c) valid   d) valid

4. a) valid   b) valid   c) valid   d) not valid

5. a) false   b) true   c) true   d) true

6. a) You will get good grades; valid

   b) You did not order the computer online. : valid

   c) He does not go swimming. : not valid

   d) I will pay my tuition. ; valid

   e) You did not sell the house. ; valid

   f) You receive your salary. ; not valid

Exercise 6

1. a) You do not go out at 12 p.m. or you will miss lunch.

   b) If she is cooking, then she is washing the dishes.

   c) She will play games often or she will study.

   d) If Afua has not retired, then she is still working.

   e) The air conditioner is not on or the room is warm.

   f) If Kojo receives his salary, then he will buy me a gift.

2. a) If he cannot apply for this job, then he is not a college graduate.

   b) If I do not buy her a car, then she does not pass the bar exam.

   c) If the dog barks, then the man is at the gate.

   d) If I travel, then the bus is not overbooked.

   e) If she is a singer, then she is well known.

3. a) If he gets a job at the factory, then he will buy a car.

   b) If he enters the competition, then he will win the top prize.

   c) If he go to college, then he will go on tour.

   d) If I do not write an article, then I will have dinner at the restaurant.

   e) If I buy the shirt, then I will have lunch.

Exercise 7

1. Valid   2. Valid   3. Not valid   4. valid

5. Not valid   6. Valid   7. Valid   8. Not valid

Review Exercises

1. a) statement   b) not   c) statement

   d) statement   e) not

2. a) compound, conjunction   b) simple

   c) compound, disjunction   d) compound, conditional

   e) simple   f) compound, biconditional

3. a) I will not study hard.

   b) Esi is not talented.

   c) I will not wash my car.

   d) Afua does not do her homework.

4. a)

p	q	~q	q ∨ ~q	p ∧ (q ∨ ~q)
T	T	F	T	T
T	F	T	T	T
F	T	F	T	F
F	F	T	T	F

   b)

p	q	~q	p ∨ q	p ∧ ~q	(p ∨ q) ∧ (p ∧ ~q)
T	T	F	T	F	F
T	F	T	T	T	T
F	T	F	T	F	F
F	F	T	F	F	F

c)

p	q	~p	~q	p → q	(p → q) ∧ ~q	[(p → q) ∧ ~q] → ~p
T	T	F	F	T	F	T
T	F	F	T	F	T	F
F	T	T	F	T	F	T
F	F	T	T	T	T	T

d)

p	q	~p	~q	p ∧ q	~p ∧ ~q	(p ∧ q) ∨ (~p ∧ ~q)
T	T	F	F	T	F	T
T	F	F	T	F	F	F
F	T	T	F	F	F	F
F	F	T	T	F	T	T

5. a) p: I will study.

    q: I will go fishing.

       p ∨ q

   b) p: The boy is intelligent.

    q: He will make a good grade.

       p ∧ q

c) p: I will dance.

    q: I will go fishing.

       p ↔ q

   d) p: I study hard.

    q; I will pass the text.

       p → q

6. a) Converse: If it is raining, then it is cloudy.

Inverse: If it is not cloudy, then it is not raining.

Contrapositive: If it is not raining, then it is not cloudy.

b) Converse: If I miss the program, then I will walk home.

Inverse: If I do not walk home, I will not miss the program.

Contrapositive: If I do not miss the program, then I do not walk home.

c) Converse: If Esi is expelled, the she does not obey school rules.

Inverse: If Esi obeys school rules, then she will not be expelled.

Contrapositive: If Esi is not expel, then she obeys school rules.

d) Converse: If I do not go fishing, then I will go to school.

Inverse: If I do not go to school, then I will go fishing.

Contrapositive: If I go fishing, then I will not go to school.

7. a) equivalent   b) not   c) equivalent

   d) equivalent   e) equivalent   f) not

8. a) not valid   b) valid   c) not valid

   d) valid   e) valid

9. a) valid   b) valid   c) not valid

   d) valid   e) valid

10. a) true   b) not true   c) true

    d) not true   e) true

12. a) valid   b) not valid   c) valid

    d) valid   e) not valid